Disaster Mitigation

Disaster Mitigation

Prof. Ashok Kumar Sharma

GenNext Publication

ISBN : 978-93-8022-277-6

Price!;!₹ 100

First Published, 2015

Published by :

Gennext Publications
5, Ansari Road, Dariyaganj,
New Delhi-110002 (India)
Mob. : 9811692060
E-mail: gennext@hotmail.com

Printed at : G. Print Process, Delhi

Cataloging in Publication Data—DK

Courtesy: D.K. Agencies (P) Ltd. <docinfo@dkagencies.com>

Sharma, Ashok Kumar *(Writer on disaster)*, **author.**

Disaster mitigation / Ashok Kumar Sharma.

pages cm

ISBN 9789380222776

1. Emergency management—Asia. 2. Emergency management. 3. Hazard mitigation—Asia. 4. Hazard mitigation. I. Title.

DDC 363.346095 23

Disaster Mitigation

Disaster is a word found in Oxford Dictionary derived from French word *'Desastre'*. Disaster means negation or destruction, mitigation means to reduce the effects of disaster. Disaster may be caused as an after effect of natural calamity or manmade calamity or as an effect of combination of both to effect atmosphere, biosphere or hydrosphere. It may be in the forms Volcanic eruptions or Avalanche, Landslide Amaranth (mudslide/Rockfall) Transportation (Air, Water, Rail, Road), climatic (draught famine) Epidemic, Plague (Surat Gujarat) Earthquake (Maharashtra Latur, Bhuj Ahemdabad) fire (Ankleshwar, Oil drilling deep sea platform accident by a boat setting ablaze the platform), Brahamputra flooding or Jharia coal mines, Japan, Indonesia, Phillifines or Jwalamukhi in India.

The Second 2nd Wednesday of October every year, is announced as International Day for Disaster Mitigation world over.

Disasters have existed ever since the existence of mankind and no community is immune to the emergencies caused by natural and man made disaster. Worldwide natural disasters have been known to be one of the major problems in terms of mortality, number of people adversely affected and economic losses. There is no evidence to suggest that frequency of natural disasters will alter significantly in immediate future. The Natural disaster has tended to be more destructive in the last few decades as they affect a larger concentration of population, due to increased urbanization and settlements in disaster prone areas.

The spectrum of occurrence of disasters indicates that Asian Region is one of the most disaster prone regions as 60% of the major natural disasters reported in the world occur in this region. India is amongst the most disaster prone countries in the world due to high vulnerability to natural disasters like floods, earthquakes, cyclones and droughts. In India flood affects over 9 million hectares area annually, 56% of landmass is vulnerable to seismic activity of varying degree and 5700 kms long coastline is prone to severe cyclone with very severe loss of life and economic damage. Besides the natural disaster, India is also vulnerable to manmade disasters like transportation accidents, chemical and technological disasters and other such events.

It is evident form spectrum of occurrence that adequate procedures to deal with disaster situations are necessary. Disaster management requires well-coordinated public policy for disaster prevention, mitigation, preparedness, emergency response and reconstruction. Appropriate plans are necessary as the disaster implicates many resource organisation at various stages, which makes it necessary to have coordination. Health care in disaster is one of the critical elements.

Disaster Defined

The concept and definition of disaster has altered over times in accordance with changing concepts concerning their cause and effect. The definition of disaster has reflected this change, with increasing attention being given to the social aspects of disaster situations and collective ability to meet the requirements of the situation. disaster has been defined in various ways on the basis of degree of physical impact of the event, magnitude, disruption of public safety, destruction of resources and in terms of special efforts required and controllability of events.

Pan American Health Organisation has described Disaster as an overwhelming ecological disruption occurring on a scale sufficient to require outside assistance. Sociologically "A disaster is an event in time and space which produces the condition whereby the continuity of the structure processes of social units become problematic."

Commonly disasters are defined as an 'overwhelming ecological disruption which exceeds the capacity of a community to adjust and consequently requires assistance from the outside. W Nick Carter defined it as 'an event, natural or manmade, sudden or progressive, which impacts with such severity that the affected community has to respond by taking exceptional measures.

Ignoring Warning Signals

Science has developed to great extent but still it cannot match the mood and power of the nature. The science may give warnings prior to the onset of any disaster but even if its accuracy is a matter of debate, less attention is paid to heed to the signal. The apathy or neglect on part of the people is always dangerous to sustain life. and is always painful. Though Abraham Lincoln said this in context of American elections but it holds good in other spheres of life too; Elections are like a fire, if you turn your back to it, you end up with blisters which are painful to sit on for next five years. Did the people pay heed to the warning and did the poll percentage improve any further, if no, then why to repent. Those who can not remember the past are prone to repeat it but it is an innate tendency of the mankind that one goes to the funeral ground and remains thoughtful till he is there, once he comes out he forgets all and becomes the same old self. Poet Shelley has said that effects of our surroundings are so powerful that the till such time we are alive we remain colorful and ignore warnings;

Life like a dome of coloured glass
Stains the white radiance of eternity,
Till death fragments it to pieces.

It is the fury of nature, which can not be matched by man. It strides the surroundings with outrageous results due to volcanoes and earthquakes They cause changes in existing Landforms. It is a wrong notion to say that history can be made but not the geography; earthquakes can alter Geography, and boundaries. They may cause division to open up in the grounds which may lead to landslides or cause giant waves in the oceans. Further to add under water atomic nuclear tests are dangerous experiments and are to be completely forbidden by G-8 summits. Beside the tremors of tests, nuclear radiations have badly affected the Japanese coasts. It is difficult to forget 4th & 5th Aug. 1945 during the second world war when America had dropped two Atom bombs on Hiroshima and Nagasaki in Japan. The sea life, fauna & mankind are still suffering from its after effects. Many children were still born and deformed due to the nuclear bomb explosion. It's a curse of mankind.

Classification of Disaster

Disaster had, has and shall affect us. We have to live with it except that the warning signals should be taken seriously. We should get ready to learn from our follies and have a better way of life. Disaster is classified into Three types:

1. Natural Disaster.
2. Anthropogenic origin of Disaster
3. The Hybrid type

1. Natural Disaster

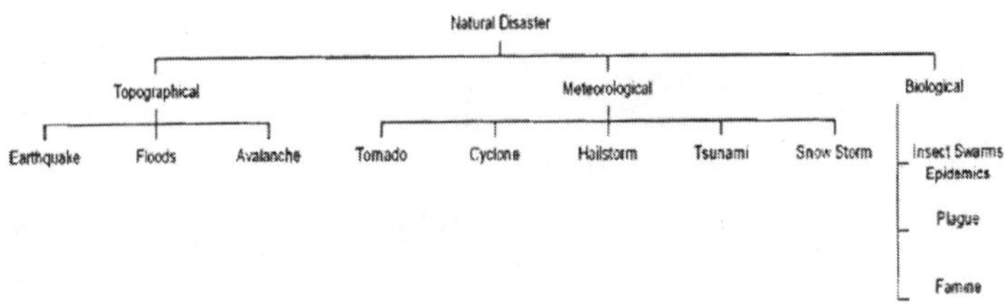

Earthquake

Peru has become the latest country to feel the renewed heat from the RING — OF — FIRE that unleashes earthquakes around the pacific almost every day. Atleast 337 people were killed after the 7.9 magnitude quake rattled the country on October' 2007.

The RING — OF FIRE stretches along the western coast of the America's through the Island nations of the South Pacific and on through South East Asia.

It is a series of fault lines in the hardened upper layers of the earth's crust. These lines of weakness are the meeting points of huge continental plates that make up the crust and which literally floats on the molten rock of the earth's core. These tectonic plates are in constant motion clashing into each other or moving away from each other, creating stresses and pressure built up at their margins.

This stress is released through volcanic eruptions, when the molten rock is ejected as magma through fissures in the crust or via earthquakes, when the pressure causes the crust to buckle and shift.

Earthquake in Latur & Bhuj on 26th Jan 2001 in Gujarat caused great damage killing more that 20,000 people and destroying 4,00, 000 houses, 16 million people were affected by its impact. Alike G-8 Summit, individually, the government of Gujarat should pay more heed to the warnings and allow construction of quake-resistant houses in future. The international agencies in their report have said that more than developed countries it is the developing countries which suffer more during disaster because they suffer from financial resources to expand hence they cluster together causing influx of population growth at limited places. Greater the influxes of population at one place more are the chances for suffering.

There are many religious and political leaders in India who are against Family planning. They argue that children are the wish of GOD & we should obey it. Political leaders argue in favour of more hands to a poor family to assist them in work and raise money. Both arguments are against natural logic to develop national GDP, and avoid clustering habitat.

Volcanoes

Proofs are being dug out to confirm that giant volcanic eruptions had caused the mass extinction of Dinosaurs between 63 to 67 million years ago. Some believe, it was an asteroid or Comet impact which left a vast crater at Chic-Xulub on the coast of Mexico that resulted in the Cretaceous-tertiary (K-T) extinction to wipe-off all dinosaurs. Others consider that a series of heavy volcanic eruptions created the huge Deccan traps lava bed in India, whose original extent may have covered as much as 1.5 Million square Kilometers or more than twice the area of Texas.

Miss Gerta Keller, USA, suggested in a Paper that the mass extinction happened at or just after the biggest phase of the Deccan eruptions, which spewed 80 per cent of the lava found at the Deccan traps. Remains of a microscopic shell forming organisms known as Foraminifera were found. The extinction event after Deccan eruptions is believed to be within 3 lacs to 5 lac years and before the mass extinction period most of the Foraminifera species were comparatively large, very flamboyant, very specialized, very ornate, with many chambers. These Foraminifera were roughly 200 — 350 microns large or a fifth to a third of a millimeter long. During the period of cretaceous tertiary extinction or K-T event the environment had changed which prompted the extinction of Foraminifera to a very tiny, one-twentieth the size of species before, with absolutely no ornamentation, just a few chambers. This hints that these eruptions came before the mass extinction and might have caused it.

The Dinosaurs might have faced a co-incidence of extinction of Deccan volcanic punch, rather then leash hit by an asteroid or comet from the space. It was a Deccan eruption which caused the significant impact.

A recent theory in favor of extinction of Dinosaurs says it was the bacterial Infection similar to Malarial parasite which has caused G.I. tract infection amongst Dinosaurs and it's extinction from the globe.

Mount Etna in Sicily, Kilimanjaro in Tanzania or Kilauea and Mauna Kea in Hawaii are examples of Volcanoes. A volcano is a vent or an opening at a weak spot in the earth crust through which magma erupts onto the surface as Lava. The Island of Curtsey near Iceland is a volcanic Island.

Cyclone

During the first week of JUNE 2007 Cyclone struck the desert Muscat. Cyclone Gonu with winds unto 170 Km/hr. slammed into Omans eastern coast and waves unto 12 Mts. high lashed out at the coastal towns.

It was the gulfs fiercest storm in 30 years. Palm fronds found in Maldivian Island were used to make a floating bag which saved 40,000 lives in Oman and Iran hence this cyclone is called Cyclone Gonu.

Anthropogenic Origin of Disaster

The Anthropogenic disasters are consequence to man's folly with the artificial surroundings which he makes. The TITANIC ship had a tag on non destructible Creation by mankind. It had all the luxuries of life on it and was considered as most safe ship to travel in high seas.

We may call it the agony of fate that it met most deadly icebergs in North Atlantic Ocean and was broken into two halves. More than 1200 people sank with the ship and perished to leave tales behind them. Take for example the inaptitude's of the ground staff who failed to close the baggage doors of Turkish DC-10 airliner which crashed outside the Paris Airport.

It was a big human failing which resulted into disaster. One can think of the monster when it is out of the bag. Can one forget the nuclear accident of Chernobyl or tragedy of Union Carbide gas leak at Bhopal?

Let us consider another example of construction of Under Sea Restaurant at Maldives. This again is an adventure in the sea of which a time will only tell. In Maldives 15th April marks the day that the first ever all-glass undersea restaurant in the world opens its doors for business at the Hilton Maldives Resort & Spa. Ithaa* will sit five meters below the waves of the Indian Ocean, surrounded by a vibrant coral reef and encased in clear acrylic offering diners 270-degrees of panoramic underwater views.

"We have used aquarium technology to put diners face-to-face with the stunning underwater environment of the Maldives", says Carsten Schieck, General Manager of Hilton Maldives Resorts & Spa. "Our guests always comment on being blown away by the colour, clarity, and beauty of the underwater world in the Maldives, so it seemed the perfect idea to build a restaurant where diners can experience fine cuisine and take time to enjoy the views-without ever getting their feet wet."

Created by MJ Murphy Ltd, a design consultancy based in New Zealand, Ithaa's distinctive feature is the use of curved transparent acrylic walls and roof, similar to those used in aquarium attractions.

"The fact that the entire restaurant except for the floor is made of clear acrylic makes this unique in the world," continues Schieck, "We are currently planting a coral reef garden to add to the spectacular views of the rays, sharks and many colourful fish that live around the area.

Fig. Ithaa-Maldness undersea Restaurant.

The Hybrid Disaster

The combination of Natural and man made events result into the Hybrid type disaster. Surat is a big but dirty town in Gujarat. The inhabitants and the government agencies were responsible to pile garbage in the town. The garbage attracted lot of Rats to found fodder to eat. Many rats died & it resulted into an epidemic of plague. People were travelling from one township to the other as carriers of the disease with fear to spreading of Epidemic amongst the community, which had no immunity to resist plague. Another example is Smog of London, which was caused by coal burning to result in respiratory diseases and pulmonary failures and deaths. Subsequently, the London metropolis framed legal laws not to burn coal. Shri Sundar Lal Bahuguna in Tehri, Gharwal, was against cutting of trees in Tehri. His followers were advised to embrace the Trunk of the trees. This movement became famous as Chipko Andolan. Destruction of rain forests causes reduction of evapro-transpiration, which could have intensified occurrence of floods. Drainage of oil into the ocean by tankers result into environment degradation and is disastrous for sea life and mankind.

A Massive Challenge

Our environment is God's gift to us. We have neither made it nor have we maintained it but we definitely have the control to inherit it. In these times of global warming we have to let our children see nature in its real form. The latest example (21/6/2007) of combination of natural and manmade events in China is building two new power stations a week to cause excess CO2 emission and to enhance global warming. A Netherlands Environmental Assessment Agency suggested; China might be the world's greatest polluter with emissions rising by 9% in 2006 compared with 1.4% in America. It is a massive challenge who will respond to it? It is only China which has to choose a pathway to prosperity while protecting the climate, morally.

Can the change in climate in one country be of any financial importance to the other country? The answer is yes! The severe winter in Europe means good income for India. Severe winter will obviously require more consumption of oil and gas for heating which in turn will enhance the environmental green house gases.

Gondwanaland

Besides, volcanic was formed islands there is a belief that the Himalayan and Tibetan plateau when a Super-continent broke up and Indian subcontinent smashed at high speed into Eurasia as reported in British Journal — *Nature.*

Until 140 million years ago, modern day India formed part of super continent called GONDWANALAND, which broke apart from what today is Africa, India, Australia, South America and Antarctica, it says that India moved at much higher speed than other parts at between 18cm — 20 cm per year.

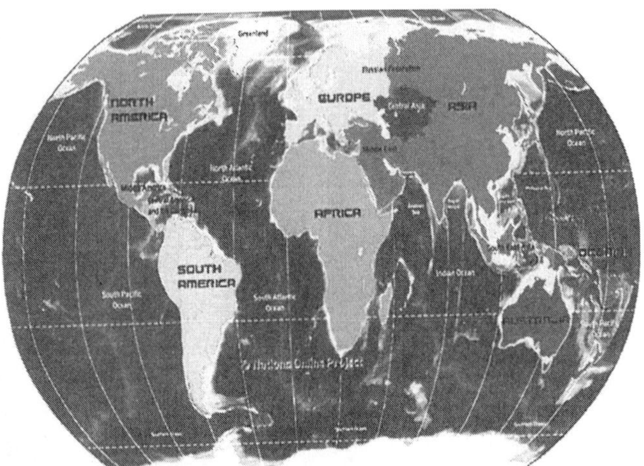

By comparison Australia, Africa plates moved at just 4 cm per year and also travelled much smaller distances while Antarctica remained almost stationary. Zipping upto north, India smashed into the continent of Eurasia about 50 million years ago and the mighty collision caused land to crumple and rear up creating the world's largest mountain range — Himalayas.

Tectonic Plates (Tekton)

The lithosphere of the earth is broken into large Jigsaw shape. These Zigsaws are known as Plates,These plates float on ASTHENOSPHEREs and move slowly by moving teeth against each other theirby pushing and pulling Sliding, rising, sinking, bending, breaking thus producing vibrations. This process is described as tectonic process. Tectonic word is derived from Greek word "Tecton" which means "to build". The seismic waves are produced at the margin of jigsaw moving end-plates of lithosphere which is responsible to cause earthquakes and volcanic activities. An earthquake leads to form fold mountains such as Himalayas in Asia, Rockies in North America and Andies in South America.

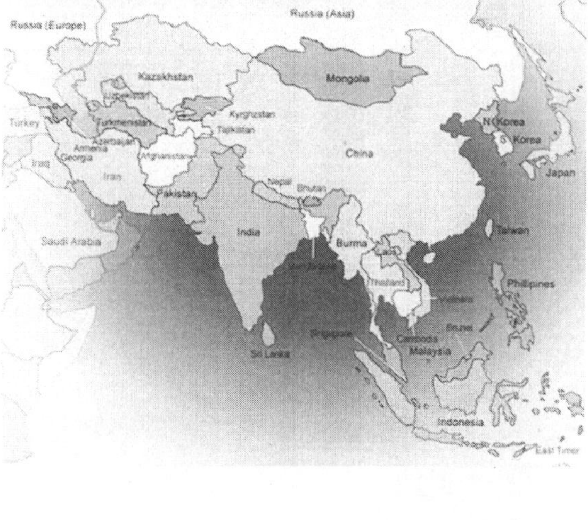

Layers of the Earth

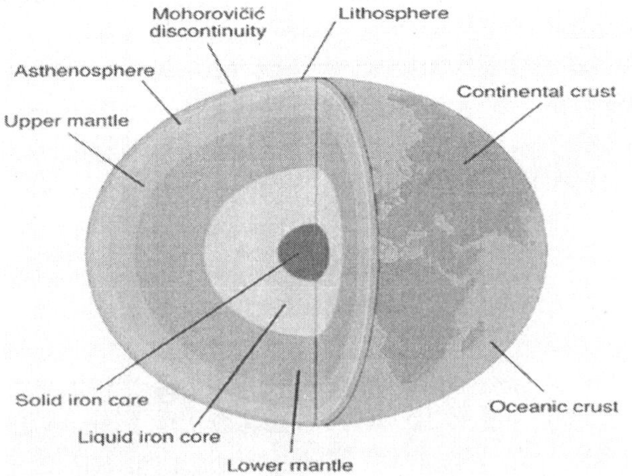

The radius of earth sphere is 6,400 Kms. Earthquakes have taught us to measure the epicenter and structure of earth. Earth movements produce vibrations known as seismic waves. The vibrations travel through different crusts of the earth with different speeds.

When they pass from one crust to another the seismic waves are bent. The speed and the bending of the waves give information regarding the arrangements of the material through which it passes i.e. crust, mantle and the core.

Theory of Formation of Indian Sub-Continent

(By GeoForschungs Zentrum, GFZ)

It is believed that thickness of Indian lithosphere is 100 km as measured by German's National Geo-Science lab. at

Pots Dam known as GeoForschungs Zentrum (GFZ) obtained through shear wave receiver function technique. The other parts of the Gondwainaland as compared with India, the thickness of their lithosphere is estimated as 180-300 km and the roots of their lithosphere have spread into the mantle. This limits the ability of the other sub-continents either to remain stationary like Antarctica or move slowly as Africa or American Sub-continent.

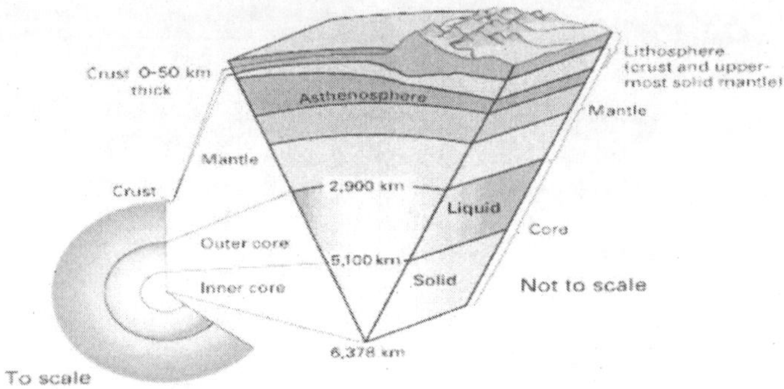

The GFZ, Germany explains that after a severe earthquake, the Gondwanaland had bifurcated into two parts and it is believed that Indian Sub-continent may have been on the top or near the fractured earth . Plume had spewed from fractured mantle to the top of lithosphere and the lithosphere roots embedded in the mantle were burnt out due to intense heat generated from the mantle. Presently, the sub-continent of Indian plate rests on a fluid and floats roam and moves in the words of Geo-scientists of Sydney University, Dietmar Muellar like "a cheetah".

In Hindu mythology, it is believed that the earth rests on the Fins of Sheshnaga which is the symbol of hot elastic mantle layer and the bed of the Sheshnaga is described as Chhir Sagar which is scorching rich core synonymous with oceanic crust. The movement of the Sheshnaga's head and neck is responsible for the movement of the earth. The bed of lord Vishnu similarly has been described as Sheshnaga.

During the Shakespearean era the medical science had not known that the circulation of blood in human beings after being oxygenated from the lungs travels through the heart but Shakespeare mentioned in his writings, "You are the roody drops which circulate through the heart". The critics and literary eminent say that Shakespeare had a vision earlier than the invention as a belief which was converted into reality with the passage of time.

My intentions to explain Indian sub-continent without lithosphere roots and floating on the mantle simply resembles what the Hindu scriptures had visualized much earlier than the pre-historic period. The scriptures give into an insight to peep into the present and to think about the future as a basis of research.

U.N. Mandate - Kyoto Protocol

In the second week of December'2007, 190 countries participated to discuss UN Treaty on Climate Change. Under the UN Treaty a global deal was called the UN

Framework Convention on Climate Change known as Kyoto Protocol. The protocol demands that the 36 big industrialized countries which emit green-house gases shall reduce their emission by a fixed percentage by 2012.

The 190 odd countries at Bali discussed the scope of emission after 2012, when the Kyoto Protocol ends. Under Kyoto Protocol, at present countries which contribute for 70% of the pollution are expected to cut their carbon foot print. But, these countries wanted that after 2012, India and China due to their growing economy add to global emissions and have to cut down green-house emission.

The emission arises out of burning fossil fuels like oil and coal would put a spanner in India and China's economic growth. The Bali, convention has decided upon a new set of principles that will over the next two years help the countries decide a post 2012 deal. By 2009, the world based on Bali road map will draw up a new deal explaining the role of each country. In future, negotiations will involve transfer of clean technologies from developed to developing countries.

European countries have fixed target for emission of green house gases under Kyoto Protocol. To offset the increased emissions they would have to buy more credits from developing countries and these credits are known as carbon — credits. These carbon — credits are generated by various projects in India and China i.e. Steel, Cement manufacturers, Hostels, Hydro-electric stations and even a few temples in India. These projects invest money to adopt a cleaner technology to save on energy consumption and thereby to reduce (GHG); In exchange for Green House Gas emission for each ton of CO^2 emission avoided, the concerned project will get carbon emission certificate which they can trade in global market. So far India has generated 29 millions carbon —credits and has 139 millions further in store. The rate of each carbon — credit is 15 — 18 Euros per certificate. Ecolutions are of the opinion that within next five years the prices will go high to 35 Euros.

It is ridiculous to note that rich countries would not clean their backyards and with the money power at their command purchase carbon —credits from the poor and developing countries who have to act as their saviors.

The U.S. First global Warmer

The U.S., with just 30% of India's population emits 475% more climate change causing CO_2 from its power plants than India, simply because it produces six times more energy than India. The Indian power companies are relatively inefficient in comparison to rich country power generator.

NTPC Third Biggest Global Co/ Emitter

NTPC is the world's third biggest emitter of CO2 as far as power producers go. The name of the other Indian power companies that depend on fossil fuel for generating electricity also crop up in the list along with those of Chinese companies that seem to be fairing as bad as Indian companies.

The rich countries which continue to emit the highest per capita emissions and have played the biggest roles in pushing the planet to the edge of a crisis should extend help to developing countries like India in finding a better road to growth by financing and sharing clean technologies.

200 Million globel Warming Refugees By 2040

Global warming could create with time 200 million climate refuges by 2040 and the bulk of the affected will be people living in poor countries. The information is based on a study by German academic Jakobeit, from Hamburg University.

They have further predicted that the figure will increase 10 folds as the impact of rising temperature becomes more clearly felt. Green Peace Climatic Organisation of Germany has said; "The poorest people in world have had no hand in climate change but will be the first to suffer severely from its fallout".

James Lovelock adviser to NASA and fellow of the Royal society advocated a new hypothesis known as "Gail Hypothesis", which believed earth as a god and living being, It's a super organism of which the man is a part but not the boss of the earth. To others this Hypothesis seemed unscientific.

Methanotrophic Bacteria Vis - A -Vis Co_2 Emission

Methane is a gas which is found in marshy lands and near Geo-thermal power plants, a bacteria known as Methanotrophic has got a special affinity to grow in high acidic media and is resistant in comparison to bacteria's of its own species, eats up the methane gas and through its

Himalayan Glaciers Face Haze Threat

A huge haze of soot and other aerosols over the Indian Ocean could be playing a significant role in the melting of Himalayan glaçiers and Shivalingam described above.

It may be premature to arrive at any final conclusions since the long term studies on the above subject are under way at the Indian Institute of Tropical Meteorology (IITM), Pune. In 1992 — 2002, another study had said that a brown haze was associated with reduction in Indian Monsoon Rains fall in agricultural production, while contrary to this notion the authors of Haze theory over the Himalayas claimed that aerosol particles forming the haze are contributing to atmospheric warming as other greenhouse gases do. The Scientists contend that the haze is warming the atmosphere enough to add to the glacial melts all over the Himalayas. Many scientists in the developed countries including USA believe that haze is correlated to rural population in India and other Asian countries are responsible for global warming. They are of the opinion that the soot mostly originates from burning wood and other biomass. The noted Indian Scientists J.Srinivasan and Sulochna Gandhi had come strongly claiming that the study was faulty and exaggerated, the impact such clouds had. They have provided evidence that similar haze existed all over the world and therefore to give it the name of "Asian Brown Haze" is unique and incorrect.

Shiva Lingam At Amarnath Melts Down

The consequence of climate change has been far-reaching and clearly visible in melting of Shivalingam as shown above. The climate change is further responsible in reduction of crops yield to the line of 10% to 20% for a 1° celsius rise in temperature and 1% of G.D.P. of the nation is devoted to mitigate the effects of climate change.

(Rakesh Tuli, director, NBRI, Lucknow, dt.20/07/07 in INSA & ISEB Conference)

digestive system converts it into CO_2. The methane gas is known as natural gas or a green house gas which is twenty times more harmful than the CO_2 gas. The methane is generated when the carbonic substances are destroyed. This study has been conducted by a Professor of Biology, Peter Dunfield et. al. at University of Calgary, New Zealand and the bacteria was found near a Geo-Thermal field at Rotoray City. It is interesting to note that Kangaroos are the national animal of Australia, they are responsible to emit CO_2 by flatulence. A recent study makes Kangaroos of Australia also responsible for global warming.

Global Warming To Cause An Ice-Free Arctic Summer By 2013: -According to new NASA satellite data the Greenland's Ice sheath melted nearly 19 billion tonnes in 2007 more than the 2006 high mark, and the volume of Arctic Sea ice at Summer's end was half what it was just four years earlier. Scientists feel that Arctic Sea melting may disappear entirely by the summer of 2012. What happens in Arctic has implication for the rest of the world. Faster melting there means sea level rise and more immediate changes in winter weather because of less sea ice. UN Secretary General, Banki Moon urged there was a desperate urgency to act to curb rising seas, floods, draughts, famines and extinction of Wildlife.

SMOG

Pollution of the atmosphere has led to smog over the sky. A blanket of smog over the subcontinent is cutting down the sunlight on the earth. Fifteen to twenty years ago, the ailments of skin disease, pulmonary infections, coronary heart disease and eye disease were less in subcontinent as compared with present day scenario. The rise cannot be ascribed to greater influx of population but to 10% reduction of sunlight pure light reaching the earth. I acknowledge with thanks the findings by Padma Kumari et el. Published in geophysical research letter and as reported in the *NEW SCIENTIST* (15th Nov'07) that amount of solar radiation reaching India's land mass droped on an average by 0.86 watts per Sqmts. each year. It is the increased Industrial activity in China and India simultaneously, which has caused climatic hazard. There is an increase in smog due to multiple factors i.e. industrialization, vehicular traffic, biomass burning (Hindus burn the dead also), dust storms. Is increasing Aerosol Optical depth (AOD) due to extinction of aerosol component in the atmosphere? On the contrary, the trend is reverse in developed countries or west. There are two faces of each coin; if the Solar Dimming is the cause for health, the other face of the coin is protection of India from Global warming.

The above conference was organised by Indian National Science Academy (INSA) and International Society of Environmental Botanist (ISEB) at NBRI, Lucknow. Dr.Nityanand was of the view that there is a delicate balance of the biotic and abiotic components of the environment and all should ensure not to disturb the balance and make serious efforts to live in harmony with the environment.

Can The Smog Be Reduced

Yes! Padma Kumari says in her letter published in *new SCIENTIST* effect of smog. Meanwhile, minimum night time temperatures, without sunshine, have risen by a much greater 0.31 degrees Celsius. It is an honest step if the political parties in India and China think of GLOBAL BRIGHTENING & it is a fact that like west which too had the DIMMING of no one enjoys the Brightening of their sky. More tree plantation in near future is a good effort by Forest department & various NGOs in India. This will bring more rains to our land which in turn will cleanse the sky.

How To Create H_2

The Hydrogen can be used to substitute oil and bad burning fuel. Recent publication by engineers; Pennsylvanian University, U.S.A., have developed a new method of producing Hydrogen gas from readily available and renewable biogradable organic material. Renewable biogradable mass i.e. cellulose or glucose can be converted into Hydrogen gas for use in powering vehicles making fertilizers and treating drinking water. Presently, most of the Hydrogen is generated from non-renewable fossil fuels i.e. natural gas.

The engineers at Pennsylvania University, U.S.A., have used a method which combines electron generated bacteria and a small electrical charge in a microbial fuel cell to produce hydrogen gas. Microbial fuel cells work through the action of bacteria which can pass electrons to an anode. The electrons flow from the anode through a wire to the cathode producing an electric current. In the process, the bacteria consume organic matter in the biomass material.

Electrophoresis help generate hydrogen gas at the cathode. In routine, the above process (electro-hydro genesis) has had poor efficiency rates and low H_2 yield but the engineers have modified the process to yield 99% H_2 by chemically modifying acetic acid elements of the reactors. The acetic acid is a dead end product of glucose fermentation.

Most green house gases in the atmosphere have been put there by developing countries without having to worry about the climate. Emissions were the resultants of multiple factories in China, which caused financial growth of China. It was a contrast of interests that one country to grow was spoiling the environmental conditions of the globe for others to suffer unmindful. The report based on data on energy use and cement production reckoned China CO_2 emissions totaled 6.2 billion tonnes in 2006.The big nations now have priorities to monopolise Oil fields; world trade & world economy with least attention to pay to the greater challenges of global warming. Green Peace U.K. director Johan Sauven said; "Responsibility lies not just in Beijing but also in Washington, Brussels and Tokyo".

If no heed is paid to warnings; the retaliation is still there. The pollution like a demon would not only harm others but even the source of pollution. The Chinese pollution affected its own products i.e. toothpaste to pet food ingredients & wheat monkfish frozen Chinese juice made with unsafe colour additives. International pressure was exerted not to import Chinese goods in America. They were considered as toxic and contained melamine, blamed for dogs and cat deaths.

Loss Of Biodiversity and Change of Climate

A British said that he heard stories of Paramilitaries cutting off the arms of illiterate peasants and applying their fingerprints to land sale documents. These paramilitary personnels were actually Biofuel gangs doing atrocities on rural population for green profits.

This is amazing that on one hand green is being converted into dark and contrary to it, in America biofuel gangs are killing people for green profits. The global quest for alternative sources of environmental-friendly energy has attracted high profile support from the President George Bush and Schwarzenegger, the Governor of California. Celebrities such as Daryl Hannah, the Actress, and Willie Nelson, the country Singer are leading a campaign to promote green fuel. Yet the trend has already had disastrous consequences for tens of thousands of peasants in rural Columbia. A surge in demand for bio-fuels derived from agricultural products has unleashed a chaotic land grab by a new breed of gangster entrepreneurs hoping to cash in on the world's thirst for palm oil and related products.

Biodiversity Periods of Various Wastes

We throw various items in the waste without really knowing its period of diversity e.g. as a woolen sock thrown into the waste takes nearly one year to diversify. Leather products require microwave incinerators to dissolve otherwise, it takes 40 - 50 years to diversify in heap of garbage. The following is the list of various products which diversify over a long period of times: -

(a) Tin Canns - 100 years to diversify.

(b) Cotton bags - 5 Months

(c) Leather Shoes - 40-50 years.

(d) Plastic Bags - 1 million years.

(e) Paper bags - 1 Month.

(f) Woolen Socks - 1 year.

(g) Banana Peels - 3 — 4 Weeks.

(h) Styrofoam Cups - Eternity

Clash of Culture

All over the world tea, coffee, drinking water is sipped in Styrofoam cups of disposable nature. Have we ever thought as to where and how the used cups are disposed off, usually, they are thrown into the garbage and it remains undestroyed forever In 2006 — 07, the Railway Minister, Govt.of India, Sri Lalu Prasad Yadav, notified the use of earthen pots / *Kullarhs* for service in the railways. This was done with an aim to reduce pollution and easier to get mixed with the earth when disposed.

Pollution of our Surroundings

Heap of garbage is generated in our cities with least resources for its disposal or sewage treatment plants (STPs). Most of the sewers end up in the rivers which raises the river bed due to collection of silt. This is one of the causes for floods and remains unchecked. The river encatchment areas have been acquired by various developmental agencies for the development of townships. This enhances the chance of over flooding and loss of revenue to the state.

The cities have rag-pickers who otherwise supplement the efforts of the government and local bodies to cleanse the city just in the same manner as the rats and the rabbits are useful scavengers in agricultural field and city dwellings. Dearth of rabbits which were responsible to cleanse mosquitoes and its larvas from the agricultural fields resulted into spread of malaria. If these animals are a boon to the society, they prove to be a measure hazard for the society at large. Surat plague was caused due to accumulation of garbage in the township and subsequent infection and death of rats leading to spread of plague in Gujarat.

Besides, these animals mostly it is the Bangladeshi men, women and children who work as rag-pickers in Delhi Township. This profession though is eco-friendly but hazardous for the health. Most of the rag-pickers suffer from dermatitis, Bronchial Asthma, Diseases of the Eye, Pulmonary Tuberculosis etc. But they still compliment Delhi alone with a massive strength of 1.5 lakhs manpower. Everyday, a rag-picker picks up between 50 to 60 Kg of waste. He/She helps 1,500 — 1,800 Kg waste every month. Many NGO's have adopted it as a profession. According to their estimate, they help Municipal Corporation of Delhi save Rs.6 lacs per day but at the cost of their health.

The Earth Provides To Satisfy Everyman's Needs, But Not For Anybody's Greed

Recently in June' 2007, Russia Laid claim to North Pole

Arctic region. Russian scientists claim an underwater ridge near the North Pole. A map of the "new addition", a triangle five times the size of Britain with twice as much as oil as Saudi Arabia has been published by Russia.

The environment campaigner's said it would be disaster. Under current International Law, the countries ringing the Arctic —Russia, Canada, U.S.A., Norway and Denmark; are limited to a 200 mile Economic zone around their coasts.

A UN Convention says none can claim Jurisdiction over the Arctic seabed because the geological structure does not match the surrounding continental shelves. But Russian Scientists have returned from a 6 week mission on a nuclear ice-breaker to claim that the 1220 wide long underwater Lomonosar ridge is geologically linked to the Siberian Continental platform and similar in structure.

The region is currently administered by the International Seabed Authority, but this is now being challenged by Moscow. Experts say the ridges have 10 billion tons of gas and oil deposits and is a source of diamonds, gold, tin, manganese, nickel, lead and platinum. a World has to remember and prove what the Gandhi had said-, they have to decide between needs and greeds. Is it really the greed or population explosion that humans have domesticated the planet to such a degree that few untouched spots remain?

The greed of the man has taken him to moon and other planets. Whatever remains in the possession he cares not to protect it? Even if plenty of rains are there, there is no concern for water harvesting.

Draught

There are many places in U.P. where the threat of draught looms large. Besides dearth of interest for water harvesting at many places, many ponds have been encroached upon by land mafias. Ponds have been filled and land sold. The men have invited draught at their own and there is no reason to curse one's luck or God for the ill. Scanty rains earlier in U.P. were because of the cyclonic circulations created by low pressure area in the Bay of Bengal.

In Andhra Pradesh Experiment has been made to cause artificial rains. Sodium and Potassium salts are sprinkled by rockets or Aeroplanes in the water laden clouds to cause artificial rains to avoid draught.A proposal has been moved by UP govt. to combat draught at Bundelkhand in the same manner.

PARCHED PLANET

Cracks appear on a thirsty piece of land at Sitapur Road in the city

The state of Uttar Pradesh (2007) was in the grip of another dry spell. According to meteorological department report rainfall in U.P. had been 10% below normal from 1st June to 19th July'2007. On an average every district of the state received 256.2 mm in the given period. While in western U.P., it was 27% below normal.

Average Rainfall in U.P. (from 01" June to 19th July'2007)

S.No.	Place	Actual Rainfall (in mm)	Normal Rainfall (in mm)	Percentage
1.	Uttar Pradesh	230.5	256.2	-10

2.	Eastern U.P.	267.6	285.9	-6
3.	Western U.P.	62	221	-27
4.	Lucknow	211.9	253.7	-16
5.	Kanpur	233.5	191.7	+22
6.	Allahabad	302.5	264.3	+14
7.	Varanasi	304.2	284	+7
8.	Gorakhpur	289.4	384.4	-25

Pollution of Water

Earlier it used to be tap water which was found contaminated but as the fate would have it the water from tube wells was found contaminated as compared to sub ground level water issued the following report: -

(i) Tube well water 42% contaminated.

(ii) Sub — ground level water - 26%

(iii) Chowk , Aminabad & Hazratganj — 43.33 %

The contaminated water is responsible for Cholera, G.I.Tract Choliform and disturbances, Typhoid. The faecal contamination water is assessed with one faecal choliform per 100 ml of water.

Ground Water is not Safe Either : - Officials of the Central Pollution Control Board are of the opinion that a few new colonies of New Delhi have high level of ground water but it is poor in quality; it contains high percentage of Iron in the form of Ferrous which when gets in contact with air converts to Ferric which is insoluble and makes the water brown, such water cannot be used for cooking or bathing and is bad for the skin. The ground water in more than over 30% areas in Delhi has fluoride content which is more than the prescribed limit of 1.5mgm/litre. The water here is saline in nature. At shallow levels the concentration of nitrate in ground water collected from domestic sewage, livestock rearing, landfills run off from fertilized fields and unlined drains is between 1 mgm/litre to 945 mgm/litre.

Major contributors to the presence of fluoride in ground water are use of fluoride salts in steel, aluminum, brick and tile industries. Agricultural discharge also contributes in a big way. Improper use of soak pits and absence of central STPs in many parts of Delhi are the other sources for pollution of water in Delhi. Poor maintenance of hand pumps and unhygienic conditions around the structure are further responsible for presence of choliform and faecal-choliform level of bacteria in water drawn from the hand pumps.

Ground water and water flowing from taps in Delhi are contaminated to cause Cholera, Dysentry, Diarrhea, Jaundice and Typhoid etc. Health effects of land are more severe in children as it can retard their growth and impact their mental development. In adults, the lead poisoning can lead to kidney problems and hypertension, excess of Iron in bathing water may cause hair loss and skin allergies; high doses of Chlorine in the water can lead to corrosion of the G.I.tract and skin allergies further a link has been found between Chlorine and heart diseases and cancer. The content of ammonia in Yamuna water is also alarming.

Role of Water Treatment Industry

The concept of water harvesting and watershed development is in circulation in many metro cities of India. Town development agencies do not sanction any new construction plans without inclusion of the provisions of water harvesting. This is a healthy and wise decision. Still more is expected from water treatment industry. It is expected that they will take more precious initiatives in developing cost-effective technologies for conservation of water.

Pollution of Air

The ITRC, Lucknow, conducted a survey in Lucknow to detect pollution of air in August' 2007. The report of the survey conducted is illustrated below by courtesy ITRC, Lucknow. Emission of smoke by cars, aeroplanes, factories have badly affected and polluted the air in the city. The contents of SPM and RSPM have reached peak level in residential colonies. Less than 10mm circumference inhaled to enter the lungs is sufficient to cause Cardio-Pulmonary diseases. The following chart gives the details of pollution in air in different localities of the Lucknow town.

S. No.	Place	SPMM gm/cu.mts	RSPN Mgm/cu.mts	Sulphur Gas	Nitrogen Gas	Remarks
1.	Aliganj	338.7	172.7	18.4	27.2	Survey
2.	Vikash Nagar	312.0	178.0	22.6	23.1	done by
3.	Indira Nagar	348.4	185.0	23.2	30.0	courtesy
4.	Gomti Nagar	336.4	164.9	20.6	23.6	I.T.R.C.,
5.	Hussainabad	419.7	220.7	31.7	37.7	Luck-
6.	Charbagh	390.2	196.3	29.7	40.1	now

7.	Alambagh	350.1	188.6	23.9	34.8
8.	Aminabad	347.2	173.6	17.5	21.1
9.	Chowk	335.9	184.7	22.3	33.2
10.	Amausi	317.1	157.0	14.4	18.5

Thirty Dirty, A Curse of Growing Economy

Blacksmith Institute of USA reports that out of thirty most polluted spots in the world, two spots are located in India. One spot is Sukinda valley (4) in Odisha and the other is Vapi(5) in Gujarat. The thirty dirty spots world over are closely linked with fast growing economies, the following is the chart of tainted ten as shown below: -

S.No.	City	Country	Remarks
1.	Sumgayit	Azerbaijan	
2.	Linfen	China	
3.	Tainying	China	
4.	Sukinda	Odisha, India	The air and soil both are polluted. In Sukinda, 70% of surface water and 60% of drinking water contain Hexavalent Chromium.
5.	Vapi	Gujarat, India	It has 50 Industrial houses and more than 1000 independent units.
6.	La oroya	Peru	
7.	Ozerzhink	Russia	
8.	Norilsk	Russia	
9.	Chemobyl	Ukraine	
10.	Kabwe	Zambia	

Till 2006, neither Odisha nor Gujarat were in the tainted list. Sukinda has the largest Chromite ore deposits in India and further it has the largest opencast Chromite mine in the world.

Approximately, 70% of the surface water and 60% of the drinking water contain hexavalent chromium caused by the dumping of waste products. The air and soil both are polluted; nine out of ten deaths occurring near mining are chromite related. Nearby, 1/4th of the population is also affected by chromite poisoning. The symptoms of the disease are; (a) Gastro-intestinal tract bleeding. (b) Tuberculosis Pulmonary. (c) Bronchial Asthma. (d) Infertility. (e) Birth defects. (f) Still births, etc.

Story of Vapi

Vapi houses more than 50 industrial estates and over 1000 independent industries. The industries are mainly chemical and tanning related. The air in Vapi is polluted because of discharge of harmful oxides. The ground water contains 96 times higher mercury than the safety norms and the soil is contaminated by heavy metals. Respiratory diseases, chemical dermatitis, carcinoma, skin lung and throat cancers are commonly reported. Women suffer from miscarriage, bleeding and abnormal foetus. Mahand Industrial Estate in Maharashtra and Ranipet in Tamilnadu are the other tainted places in India. According to a survey conducted by Supreme Court Monitoring Committee, approximately, 1800 tonnes of harmful sludge has accumulated at Mahand Industrial Estate C.E.T.P. The Tamilnadu Pollution Control Board estimates 15, 00,000 tonnes of solid waste has accumulated in the town known for its tanning industry. An obvious pattern is that most of the places are located in developing world. A World Bank special study on Trade and environment, that industrialized countries share of global manufacturing exports has declined from 91% to 88%. The labour intense decline sectors are textiles apparel, footwear and other light manufacturing. Dr.Raghunandan, Secretary of Delhi Science Forum, says that most developing economies are less strict about pollution norms, therefore, polluting industries concentrate in these countries.

Ozone Hole

In 1992, Amartya Sen Nobel Prize winner stated the that depletion of Ozone Layer was due to chemical role of Chloro-Fluoro Carbons (CFC) thus creating a Ozone hole. Dr. Amartya Sen said that evolution should be left to itself which will provide a solution to the depletion of ozone layer by evolving genes that are less vulnerable to ultraviolet radiations.

Since the theory has been advocated by an eminent Nobel prize winner, it is obvious to accept that the less exposed organs/cells of the human beings when are put to task for compensation they always stand by and substitute/ compensate the actions which are lost or diminished and this is true when Dr Amartya Sen pleads evolution by itself and compensation by unexposed genes to the ultraviolet rays. But, it has certain reservations in terms of time factor responsible to evolve genetic code.

As a normal procedure genetic code is diluted in thousands of years. Dr. Sen though has based his hypothesis on firm argument, but it requires further scientific probe. Still it is desired to have a better balance between man and nature. Einstein believed man and nature are not as divorced as they seem.

Black Hole

The theory of black hole has been significantly dealt with by scientist Stephen Hawkins. It explains that when a star in the sky outlives it's life it gets dislodged from it's orbits but leaves a black hole which is not visible to the eyes but full of darkness, engulfed with heavy dust particles.

The temperature with in the hole is very high with high gravitational force. The force of the gravity is so high that it can engulf even the sun or the earth within the hole. The mass attracted towards the hole is converted into very small fragments thus converting the mass into energy. Earlier, Hawkins was of the opinion that any substance attracted inside the hole is totally consumed but Hawkins has recently changed his views to say that instead of consumption of the substance it is the transformation of substance from one form to another.

The very minute end products are filtered from one end of the hole to exit from the other end to form either a new planet or extend any existing planet.

Darwin - Science, Christian And Hindu Scriptures on Evolution: - Island of Galapagos will always be remembered due to Charles Darwin. He had advocated the theory of evolution from the same place. Two basic fundamental points i.e. evolution and creation have to be understood separately. Creator has a primary role to play which is subsequently nursed by evolution. According to Bible, the world and its inhabitants were created by God. Darwin followed later and challenged by saying that life forms change according to the requirements of their environments. The noble prize winner Dr Amartya Sen supported the concept of Darwin. Dr Sen, further explained the concept of compensation in the process of evolution which is further strengthened by today's genetic and biotechnology studies which have proved that life forms can and do mutate to adjust to their surroundings.

Soul Is The old Name for Genetic Code

Indian philosophy always believed God as the creator of all things. They believe in various Avataars at different times but in various forms in response to a particular situation or in response to the demand of their environment as propounded by Darwin. If the Gods were practicing the process of change, the rest of life on earth followed suit. The theory of Darwin and Hindu philosophy believes in the harmonious existence side by side of creation and evolution. Darwin points out that the genes of the species keep mutating through the generations according to the requirements of the surviving in a particular environment. In other words, genes are re-incarnated which explains genes to clothe with or embody in the flesh again starting from a sperm cell or an egg or grain of a pollen and as egg, the genetic code is in the flesh again. In other words, one may consider that there is incarnation and soul is the old name of genetic code.

Late Sri Prabhupad, the chief of the ISKON mission while attending a seminar on death and afterwards, held at Hotel Hilton in the city of San Francisco, USA, in early 60s was sharing the dias with Dr Christiaan Barinard of South Africa, and who had become a celebrity as the world's first hear transplant Surgeon. During the post discussion session a very pertinent question was raised when a person from the audience directed his query to Dr Burnnard asking if the heart is the abode of soul why the soul could not be detected anywhere in the heart? Still if we believe in soul and heart theory then would the doctor explain as to what characteristic changes would occur in the personality of heart recipient after transplant? Dr Barnard was perplexed to answer and he directed the query to Sri PrabhuPad ji. Prabhu ji gave an explanation that the size of the soul is so small that it cannot be seen under any microscope or through sophisticated investigations. He said, "take for example the head of a minutest needle and divided it by 1/10,000 Th micro 1.1 size which is the size of the soul. It has a virtue to stay in the heart and roam the body till six hours after the brain death of a person. A donor, who succumbs to death, for six hours the heart remains useful for transplant but his soul leaves the heart and in case of the recipient who is alive but the heart is sick, the soul remains in the heart and after the transplant the soul of the recipient enters the transplanted heart. This explains that after heart transplant the body and its organs may die but the soul remains intact. According to Hindu Scriptures, Vishnu appeared in different forms ten times,. Shiva appeared for seven times and Jesus was re — incarnated three days after the crusification and just prior to his capture, talked about his second coming. When the son of God Lord Jesus was incarnated, it was in the form of a person in tune with the accepted mode of time.

The first book called *Genesis* was by Moses, states that the Bible is perfectly in step with modern scientific discoveries when it says that first the earth was without form and void. A spirit of God was hovering over the face of the waters with potentials for creating life and it is believed that plants came before animals, in the same manner as the creatures of the Sea were created before the creatures of the Dry land; the birds were included with marine creatures and the creatures of the air came last of all.

Gaia Hypothesis (Earth as live)

Gaia Hypothesis believes earth as alive and a God. Since, it is not supported by documentary scientific evidences many don't believe in this theory, but the Hindu scriptures believe earth as a living being and as God. Ved Vyas personified earth as Kunti in the epic of Mahabharta. In Sanskrit KUM — Ti means earth. Kuntis other name is Pritha, meaning Prthvi or the Earth. Life originated on earth with the help of sun rays. Vegetation was born 'Kama' means grain with chaff. He represents seed, the essence of vegetable kingdom. A seed can not sprout without water. Kunti placed the child (Kama) in a basket and floated it in the waters of river Aswa. Kama was found and saved by Adhirath, who lived by the banks of the river. Adhirath means 'Anxiety' for body.

Karna was born bearing Kavach, alike a peel which covers the vegetable kingdom. It means a protective lining over the body. Other essentials of life are (besides sun energy) Vayu, Jal (Indra), Agni, Akash thereby meaning Panch Pandavas, which constitute the essentials of existence of man on the earth.

The above interpretation draw attention to *Panchtatvas;* the essentials of earth to constitute life and which are live. The importance of all aspects of environments for as human beings, we have within us elements of all that exists outside us. The earth (environment) and its inhabitants therefore need to live in mutual harmony. Earth is alive in this sense and has to be protected.

The Role of The UN in Complex Emergencies

(1) Creation of contingency funding arrangement.

(2) Establish a central register of specialized personnel and teams of technical specialists, supplies and other resources.

(3) Provided by a high level official, the emergency relief coordinator, designated by the secretary-general, to work with the entities of the UN system dealing with humanitarian assistance.

(4) Negotiation with all parties concerned to obtain their consent.

(5) The establishment of temporary relief corridors, days and zones of tranquility and other forms.

Reasons For Concern

(1) Displaced persons are often ineligible to receive relief and assistance available to refugees.

(2) Displaced are often insecure about relying on their own government for protection.

(3) The obstacle of national sovereignty that limits outside agencies to assist this population.

UNDRO's Mandate In Disaster Relief and Management

UNDRO has a two-fold mandate:

(1) to mobilize and co-ordinate international emergency relief to disaster stricken areas and

(2) to promote disaster preparedness and prevention activities including technical assistance to disaster-prone countries and regions.

Disaster management is a continuous process. Rehearsals to be continued to Access preparedness of Ambulances, Hospital vigil, Fire fighting equipment; management, intercommunication network etc. On the contrary, the government and public at large remains satisfied that in case of any disaster plenty of help will flow from neighboring areas and International agencies. They fail to understand that they will be beneficiary of any forthcoming help only when they will survive. The mishap always takes a big toll leaving survivors to repent.

During Latur, Bhuj and Uttarakhand earthquakes besides local help, International agencies provided tents, food packets and commodities of daily life. The help flew in abundance resulting in mismanagement. The tents were said to be sold in open market and there was loot and stampede during the distribution of water from tankers and kerosene oil. As regards post-disaster town planning is concerned, the disaster proves a boon for developmental process provided we have a vision and will to do so. If we repeat our mistakes to build our houses on landslide areas; low areas; non-earthquake resistant or of concrete in seismic zones, we are prone to suffer. Japanese have a national character and thinking that any unplanned activity will retard the progress of their loved country. It will bring political instability to the nation and the nation which has given an industrial jolt in global car market will be distracted and psyche of mankind and animals will be affected. United Nations training programme in disaster management differentiates the following zones;

To avoid the ill effects of disaster on human and animal psyche, it is essential to study the following course for remedial solutions: -

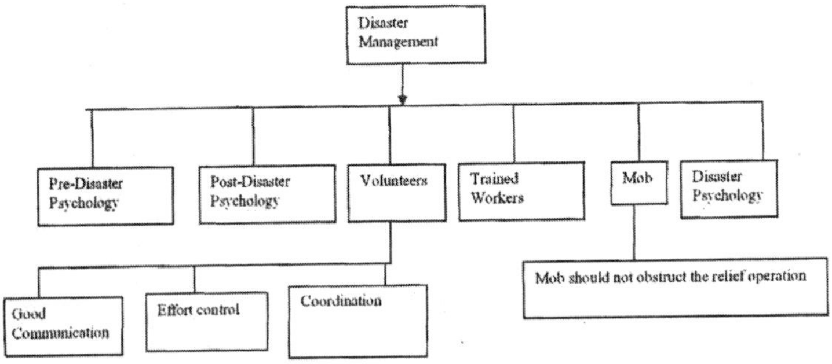

Disaster Response

The rapid disaster response is heavily depended on appropriate disaster plan and preparedness. The initial response is hinged on the rescue and relief. Rescue in specific disaster requires trained personnel and special equipment like rescue in case of earthquake to find survivors in debris/collapsed buildings. The important feature of the response includes the following :-

(a) Mechanism for repaid survey and assessment to determine immediate and medium term needs & measures required for rescue and relief.

(b) Mobilization and deployment of resources.

(i) Extrication and rescue.

(ii) On site medical care.

(aa) Mass casualty care.

(bb) Triage, First Aid, Evacuation.

(c) Transportation for definitive care at designated hospital.

(d) Hospital care:

(i) Immediate life saving.

(ii) Definitive care

(e) Long term rehabilitation.

Training for Disasters

Training is the essential aspect of the preparedness and it will determine the outcome of the response as well. Training needs attention at levels of Formation HQs, Medical units and upto the MO level. The factors to be considered are:-

(a) Creating awareness of the Risk, Vulnerability and Hazards.

(b) Dissemination of Disaster Management Plans.

(c) Awareness of role and responsibility of various functionaries.

(d) Arrangements for mobilization and its practice.

(e) Training of hospital staff in operationali zing the hospital disaster plans.

Action by MOs and NGOs

In general, the MOs and NGOs should ensure that they have adequate amount of the following items to deal with such situations. Relief officers should ensure adequate replenishment of these items, as well as evacuation of casualties, specialized treatment and coordination of resource within same and neighboring formation deployed for relief operations:-

(a) Arrest of bleeding and management of wounds, including dressing material, antiseptics, suturing material, and so on.

(b) Intravenous fluids.

(c) Resuscitation equipments, including equipments for maintenance of airway.

(d) Splints.

(e) Spraying equipment and insecticides for mosquito and fly control.

(f) Water purification, and testing equipment and chemicals, i.e. Horrocks apparatus, Chloroscope, bleaching powder.

(g) Disinfectants like sodium hypochlorite, cresol etc.

(h) Vaccines and sera, especially TT, Typhoid, and other specialized vaccines/sera as the situation demands.

(j) Drugs for first aid, management/chemoprophylaxis of common illnesses, injuries, diarrhoeal diseases, malaria, URTI, skin diseases, meningococcal diseases, adjustment problems and so on.

(k) Strtchers and necessary linen and mask and tents.

(l) Ambulance vehicle.

(m) Forms for referral, requests for air evacuation and so on.

(n) Methods for establishing communications with other functionaries.

(o) Biscuits, bread, salt, kerosene, matchbox, torch and cells, Patatoos, by NGOs.

(p) Portable ultra sound machine is a must.

Disaster Management

1. Avoid panic
2. Rapid Assessment of the situation
3. First Treat-First Casualty
4. The casualty calling for help is in less danger than a silent motionless one.
5. Priorities are ABCD i.e. Airways adequate passage, breathing is not obstructed; whether bleeding and is conscious or not ? Hold the jaw forward to maintain an open airway.
6. Request bystanders to help control traffic and mob around.
7. Motor Cyclists may have to have their helmet removed to avoid airway problem, vomiting or bleeding.
8. Gently slit along the seams of cloth to uncover the parts & look at the chest
9. Look at the colour of lips; skin (Asphyxia), Empty mouth; bleeding from nose; Ear or leakage of clear or straw coloured from the skull ;

10. Examine Pupils, whether gets smaller when light is thrown, whether blinks when Eye lashes are touched ? Whether the contour of Vertebral column at Neck is maintained or not ? Listen to the explanation of pain.
11. Whether complain of Chest pain and on inspection chest pain is present (#rib); Do not turn the patient from side to side or one position to another ; Press-Gently over the (#pelvis); position of the limbs;
12. Assess level of consciousness by :- (i) Eye opening and any movements of eye. (ii) Spoken response, touch, pain response Eye-lash reflex;
13. Check pockets, bags, in presence of a witness only and access the per cent of Burn also.

Heart Lung Resuscitation Based on ABC :

Airway:

- Tongue should not be allowed to fallback in the throat.
- Blood, Vomit or any foreign body may not obstruct the airway.

Artificial Respiration by Expired air Method

(1) Turn the casualty on his back (fig 1) kneel by his side pinch the Nose with one hand while lifting the chin forward (fig 2) with the other hand.

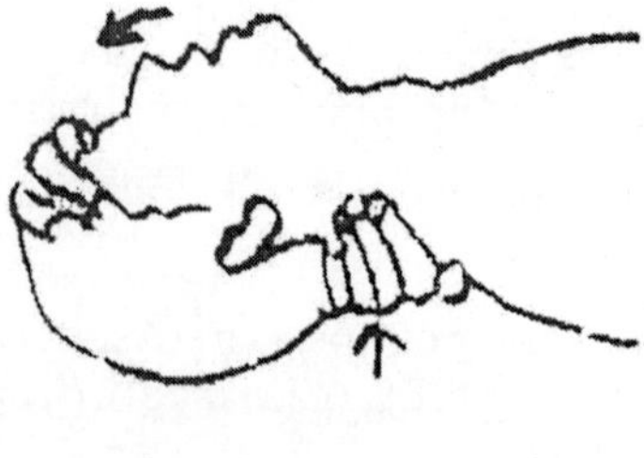

Mouth to Mouth Technique

(2) Take a long breath and place your mouth over that of the casualty. Seal your mouth with that of the casualty. In children instead mouth to nose and nose technique is used with babies.

(3) Blow to inflate casualty's chest and take your mouth onway and feel for casualty's exhaled air on your chin/nose/chin.

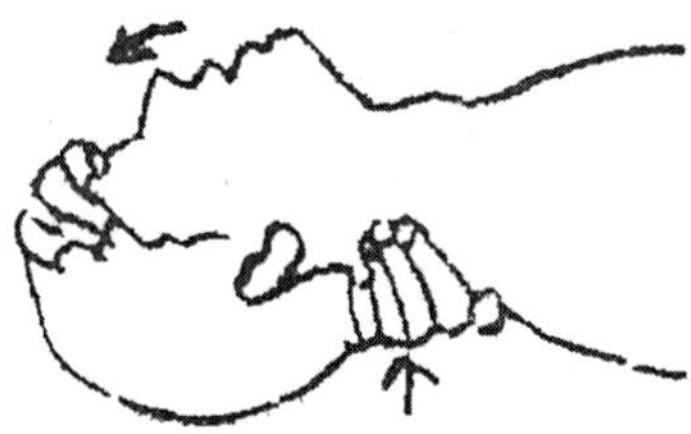

Repeat the above for (5) times and see if the casualty has started breathing again viz 10 breaths/minute (adult); 15-20/minute (children) & 25-30/minute (Babies).

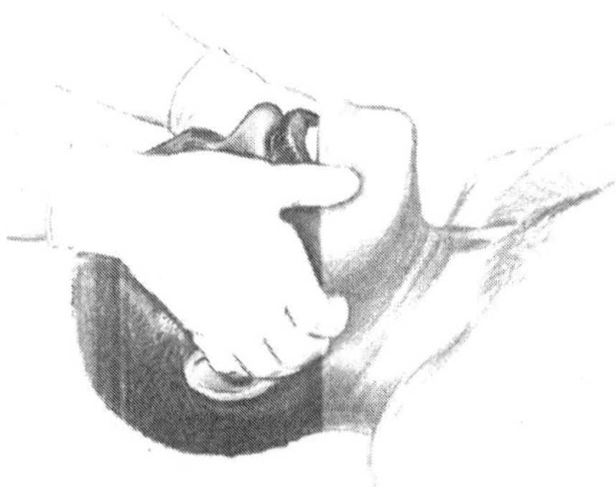

Jaw thrust without head tilt. The jaw is lifted without tilting the head. This is the airway maneuver of choice when the victim has a possible spine injury.

Watch for Cardiac Arrest following Respiratory Distress

(a) Circulation and CPR : Fall of circulation may result into pale colour of the casualty, they may be purse less; and unconscious.

CPR

(1) EAR (Expired Air Respiration) (fig 1 & 2 as above).

(2) CCCM (Closed Chest Cardiac massage) Give (5) Ear to the casualty then start chest compression. Casualty must be placed on firm surface for counter pressure. If there is a bleeding; it should be stopped simultaneously.

Locate the Notch of the sternum; place the heel of one hand over the lower 3'd of sternum and place the other hand upon the Ist rib; now the hand is over the heart. Keeping the arms straight rock forwards to compress

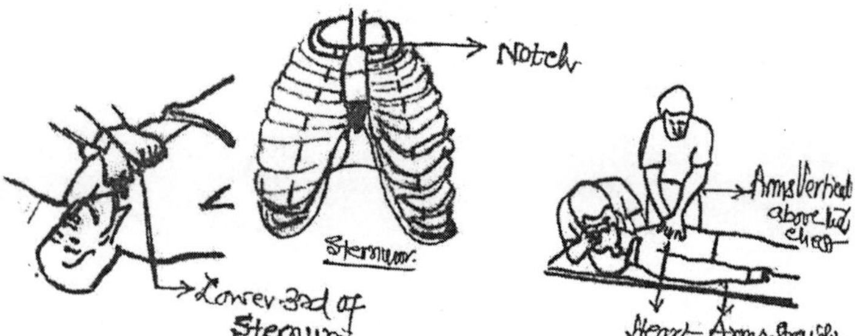

the casualty's chest about 2" (In adult). Arms should be vertically above casualty's chest. Rocking backwards and forwards alternately release and compress the heart. Children require much less pressure. The compression must not be fast, in adults the rate should be 60/minute. Even when the heart rate/beat resumes the casualty is still given artificial respiration.

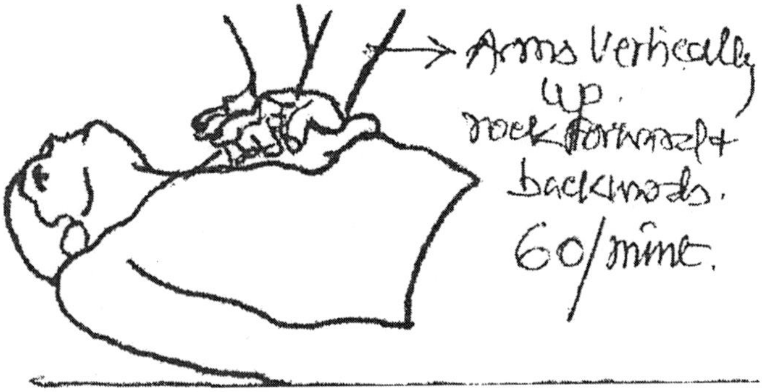

Firm Grips — A chair to lift

Precautions During the Process to Stop Bleeding

(1) Keep the patient warm.

(2) Fluid intake should be maximum.

(3) While carrying the patient Head should be lower than rest of the blood to ensure blood to the Brain.

Light bleeding :

Hemorrhage

(1) Arterial Bleeding

- It flows out in Jet form. Blood trickles out from the bandage continuously
- Apply pressure bandage. First apply your thumb on the leak firmly; than slowly relax the pressure of thumb.

Or

(2) Capillary Bleeding

It stops when dressing is applied firmly blood does not trickles continuously as a stream from the wound.

Apply Dressing.

Artenial Bleeding Dressing

Make a roll of some cotton/bandage to use as plug over the wound. Bandage the plugged wound firmly. In emergency even one's handkerchief may be made into roll by making knots one over the other to plug the wound.

Torniquests

Be careful not to apply it for very long period, to avoid irreversible soft tissue damage. Once the pressure bandage has been applied tourniquet is to be removed.

Triangular Bandage

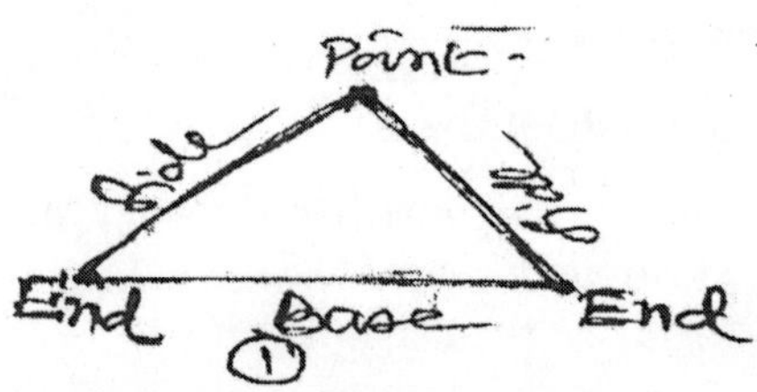

- It is applied to retain bandage or the splint in position.
- To afford support to an injured part, control bleeding, reduce pain, in following manner;

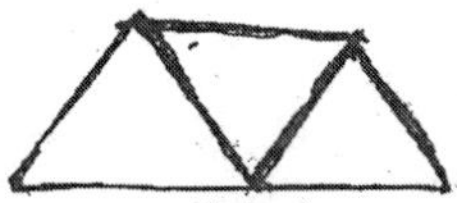

I(a)

Reef Knot-

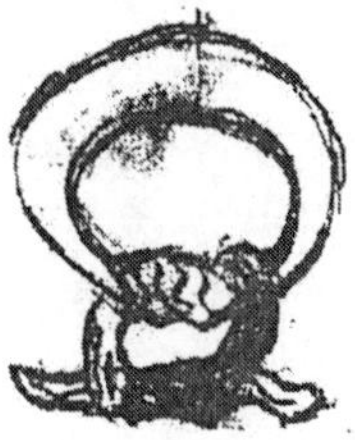

I(b)

I(c)

Burns and Scalds

(1) Do not puncture the blisters.

(2) Cover the patient with a cloth available.

(3) Keep the patient warm.

(4) Immobilise the affected part by slings out of splints.

(5) Give large quality of sweet Tea (warm fluids).

To Relieve Pain

(6) Saturate the area, including clothing with;

1. Pint of sterile water at 98.6°F (Body temp)warm

2 tsf mixed with common salt

or

Soda Bicarb

Keep the burnt area moist with above solution and apply a pad of cotton wool. In cases of bums due to cohesive chemicals (Acid Burns) bathe the part freely with an alkaline solution.

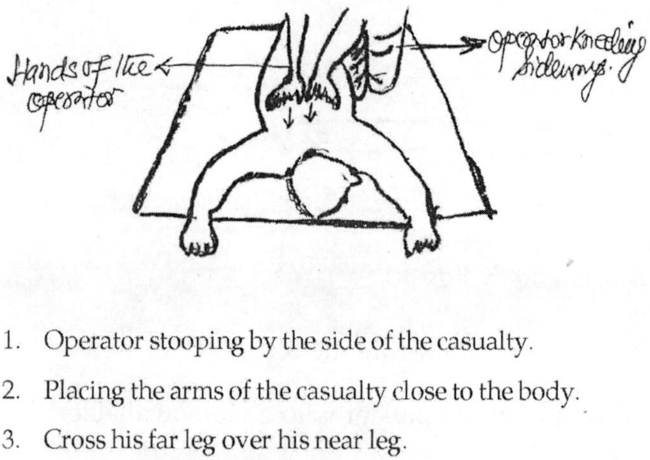

1. Operator stooping by the side of the casualty.
2. Placing the arms of the casualty close to the body.
3. Cross his far leg over his near leg.
4. Grasp his clothing at the hip on the opposite side of the body and quickly and gently turn him over while keeping your elbows quite straight and practice the following :

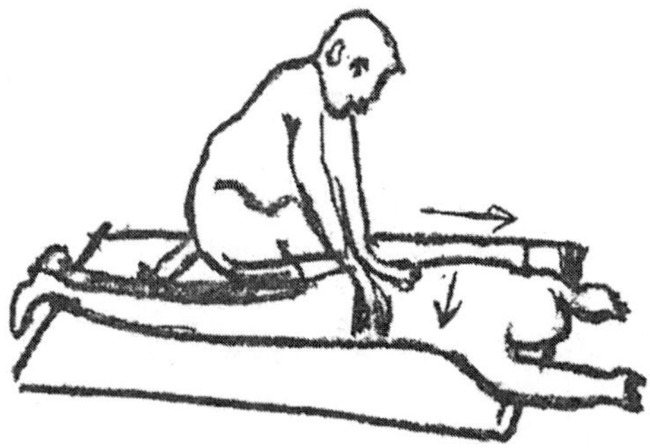

Movement (1)

Elbows straight, swing slowly forward by unbending the knees until the thighs are in an almost upright position and elbows straight allow the weight of your body to be transferred to the (posterior) pelvis.

This compresses the patients abdominal viscera to be compressed against the ground and up against the diaphragm, air is thus forced out of the lungs.

Movement (2)

Slowly swing back on to your heels and relax the compression. This causes respiration (1) & (2) should be repeated 10/minute. When natural breathing begins the rate of artificial respiration may be adopted to correspond with it.

Heat and Cold Emergencies and its Management :

Effects of Cold	Symptoms
(1) Frostbite	(1) Pain with numbness
(2) Hypothermia (30° — 35°C)	(2) Inability to move parts

(3) Gangrene	(3) Color of skin is changed to white

Treatment

(1) Try to provide shelter to the casualty. Gradual rewarming is the main aim. Do not rub affected parts, remove rings; watch and tight clothing.

(2) Warm up casualty's hand by placing them in the casualty's own armpits. Warm up feet in contact with your body. Taped water; elevation affected limbs, Do not use fires etc to heat up.

(3) Avoid windy exposure and wet clothing. No alcohol. Wrap with blanket and early removal from high altitude. Give warm drinks. Warm water bottles to trunk not limbs.

Poisons

(1) Ensure respiration and circulation. Use of resuscitation; Arrange oxygen supply or give artificial respiration.

Transport Insured

The Hummer Crutch

The Pick a- Back

The Cradle Carry

The Fire Man's lift

Two - Hand Seat

The Human Stretcher

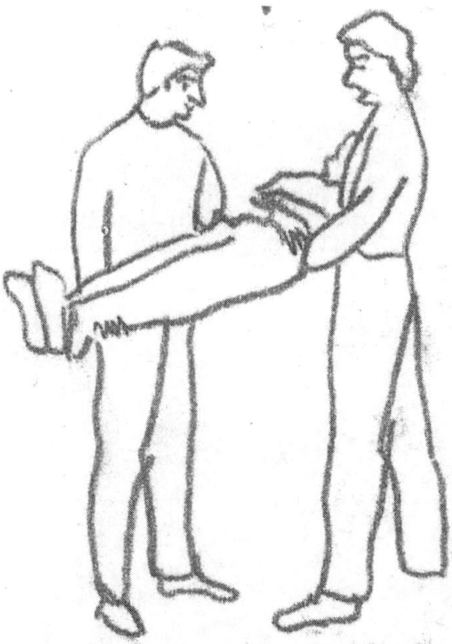

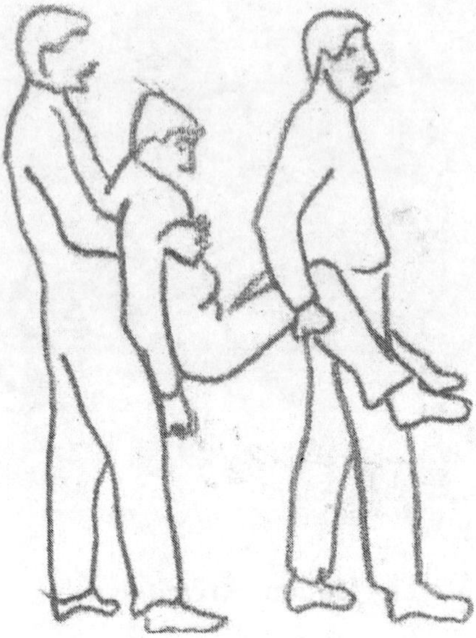

The fore and After Carry

LEVELS OF PREVENTION & REHABILITATION

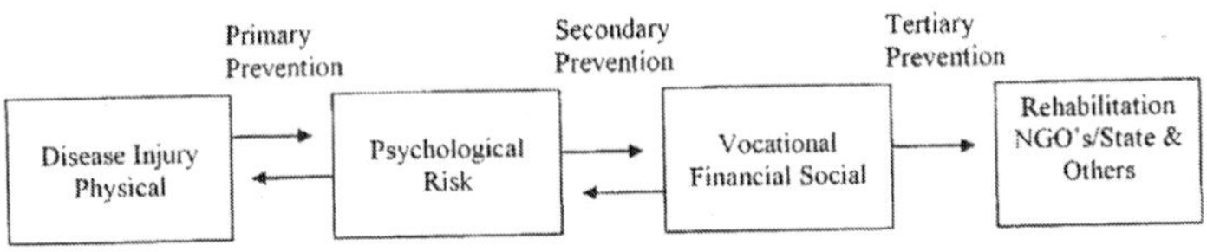

Primary Physical Prevention is based on warnings and steps taken, immunization, chlorination of drinking water. It is designed to promote general health. Habitation at risk free sites. The process requires a regular drill. Psychology risk is the factor arising out of calamity and steps to combat it. Vocational, financial and social risks are caused due to abrupt change of conditions. "

Rehabilitation is a socio-medical relief in three steps i.e. immediate attendance to save a life in peril, Nurse a life which is saved and re-installation in the social life.

Identification of the Problem

It should be,

- Simple enough so that a rapid Evaluation is possible. Do not wait for specialized investigative procedures, use clinical observation to access shock, fracture, malnutrition, internal injury (Visceral/Head), and any obstructions.

- Consistancy be maintained.
- Objective, using measurable/first aid factors.
- Describe the actual situation for near accurate reflection.
- Access immediate need to remove any obstructions, washing and clearing of wounds, light splintage/bandaging, pain relief, maintain nourishment, climate control measures, transportation to safe camp, use of mobile services to maintain communication. i.e. Broadly to categories optimally self dependent, partially independent, partially dependent and totally dependent calamity affected.
- Distribute volunteers for Site I, II, III & IV as skilled, semi skilled for attendance and referral to base camp for further management. Base camp has to decide which cases to be retained and which are to be referred to hospitals.

Site – I O.S.D.	Site – II P.I.	Site - III P.D.	Site - IV T.D.

- Distribute volunteers for Site I, II, III & IV as skilled, semi skilled for attendance and referral to base camp for further management. Base camp has to decide which cases to be retained and which are to be referred to hospitals.

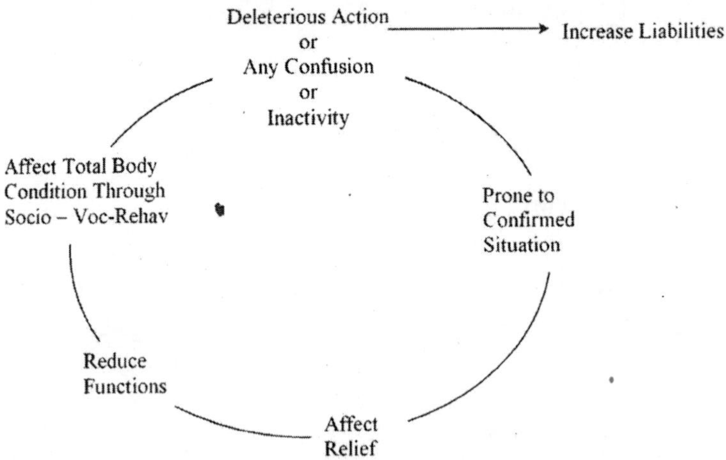

Socio - Vocational Rehabilitation

It is a team effort which starts from the site of disaster, moves to surrounding areas and ends up at the site of start. It aims at providing the calamity affected men and animals both with optimum support for reinstallation back in the main stream of the society. Vocational Rehabilitation is the concept of the right to work, a mode to dignity, an out let of aspirations to add to one's self esteem.

The NGO's provide moral support to the affected society and also creates awareness in the community about, outcome with work proposals.

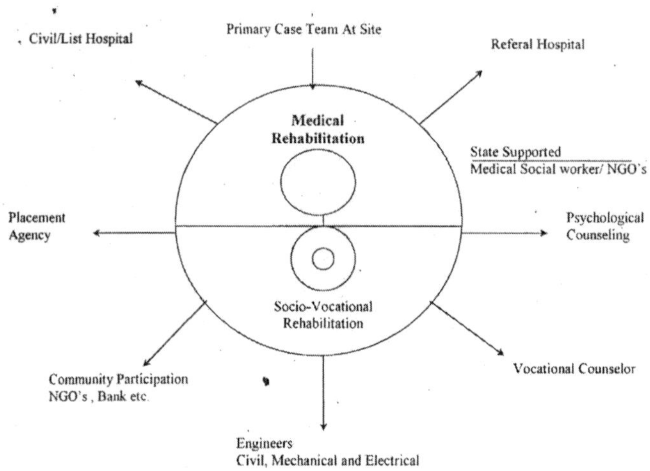

Disaster Re-Insclusion

Local Resources may be used at the site of Disaster for Rehabilitation,

1. **Walking crutch**

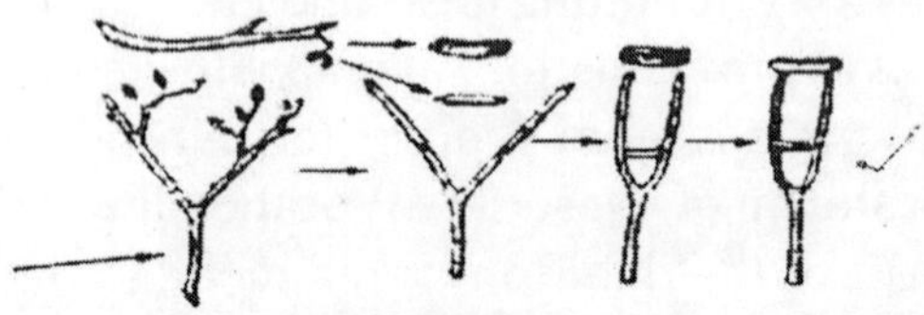

Portable Ultrasound Machine is a Must

After preliminary site management as pre-hospital care the patient is shifted to a base hospital for further specialized Emergency Care;

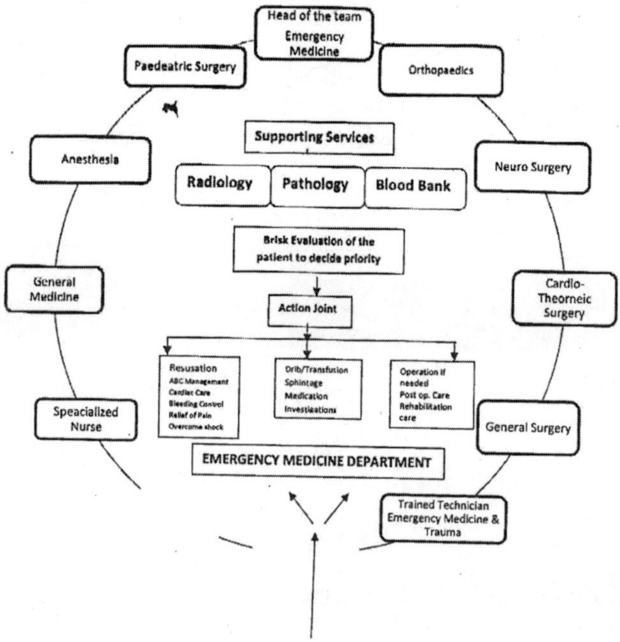

Hazard Vulnerability Analysis (HVA)

The HVA is the first step to plan disaster planning.

Mechanism of action and threats to the public health as a result of chemical events have to be understood. Post disaster complications have greater repercussions on human and animal lives. One will never forget the post disaster nuclear effects in Japan after IInd World War. Recently, "Srinagar" was flooded and life come to a stand still, once the water settled down infection and epidemics spread all over the valley. The Emergency team require understanding for,

(a) Patient delivery decision (Priority).

(b) Pre-hospital plan (Rescue operation, Emergency first aid, transportation).

(c) Clean demarcation of responsibilities of specialists, paramedical staff & ambulance staff & I/c should have knowledge to manage multi injury patients and overcome shock and administer the ABC method to ravine and bring relief of pain next step is stabilization which is done In a base camp.

(d) Long distance terrain; lack of coordination and communication; fewer resources have to be covered immediately. Establishment of communication between the site of disaster & the relief base has to be decided first.

(e) Establish protocols, induction training, good relationship between the relief team and the locals, Regular Drill of developed protocol.

Various Components to Create Concept of Emergency Medical Services

EMS evolved in 1970 in USA to reduce morbidity and mortality. It has the following contributory factors in its evolution;

(1) Public Health integration and to train community at large in first aid; Cardio-pulmonary resuscitation and knowledge of Fire law.

(2) Health care administration and organization, to train basic and advanced life support ambulances.

(3) Public safety; planning and management.

(4) Modern Medicine and surgery with its various branches and research centers, since, it has distinct entity based on ever evolving medical knowledge. The autonomy of the Emergency medical team is interdependent on multiple skills of the members of the team.

(5) Rehabilitation (Physical, Vocational, Social etc).

(6) Appropriate resource allocation to be managed through an organized approach; brotherly gesture and communication in the language of the place, for effective communications.

(7) Emergency medical technicians (EMT) should only be deployed in rescue team.

(8) "First Responder", In routine; the first responder on the scene is often the local police; BSF, CRPF etc personals or law enforcing officers of the state governments. They are non-skilled they only access the nature, cause of disaster and casualties , they are not trained in Physical rescue operations. A lot of time is wasted to complete formalities and delay rescue operations, which should be avoided.

(9) National standards for EMT (Emergency Medical Training) training should be established. A license is to be issued (Registration) by State Medical Faculties on uniform pattern of protocol.

Transportation vehicles may vary from small golf carts size to ambulances and helicopters or big transport planes fitted with Oxygen; supply system all persons involved with rescue operations should have 'First responder training'. First responded invitations and certificate of proficiency is to be issued.

All government; quasi govt; private sector or public sector, professionals have to be given F.R.T; when they join their jobs. Those who are already in the jobs it should be linked with next yearly increment or incentive. Failing to get a licentiate certificate should stop next rise. Emergency Medical Service (EMS) is responsible to cover areas of patient resource, diagnosis, transportation, field resucitation, stabilization and ultimate rehabilitation.

Disaster Management Plan of Gujarat

Gujarat state is prone to experience droughts, flash flood; Cyclones and Earthquakes. In 1978 it was morvi floods, in 1998 it was Kandla Cyclone; and in 2000 floods across Baroda, Surat, Messana and Ahmadabad & on 26th Jan' 2001 people can not forget the Earthquake of Bhuj & in 1993 in Latur.

It is very encouraging that corporate sector had joined hands with the govt. efforts to offer relief and rehabilitation besides foreign organizations and various govts at large.

World over the scene is suffering from Lack of relief measures to prevention and preparedness to mitigate disaster and to develop a mechanism to cope with knowledge to create disaster resilient, self-relient and less. vulnerable community with acquired skills to face the disaster.

Who Constitutes the Team

The Disaster Resve team includes state govt of Gujrat, govt of India, Ministry of Home Bharat Sarkar and United Nations.

It is the combined resources of the govt.; central govt., and United Nation's Disaster Programme to offer relief and rehabilitation services divided into two phases for a span of six years.

Coordination at national level is done by (PMB) Programme Management Board headed by the secretary, MHA. The Programme Steering Committee (PSc) headed by the Joint Secretary (DM) in MHA. They will meet quarterly to review the progress of the programme, at state level it is the GSDMA(Gujarat State Disaster Management Authority) as nodal agency at Gujarat state level. It executes the programme with support of UNDP. Monitoring is done by the Principal secretary and his team.

Programme Goal

Sustainable reduction in disaster risk in some of the most hazard prone districts in selected states of India.

Objectives

I National capacity building to institutionalize the system for disaster risk management in Ministry of Home Affairs, the nodal ministry for disaster management for all kinds of disasters except for drought.

II To undertake environment building, education, awareness programmes on different facets of disaster management; strengthening of capacity at all levels in disaster risk management and sustainable recovery; development of manuals and training modules; information, education and communication materials and their dissemination.

III Multi-hazard preparedness, response and mitigation plan for disaster risk management at state, district, taluka/block, village and ward level in 120 select districts of twelve most hazard prone states over a period of 6 years.

IV Networking knowledge on effective approaches, methods and tools for disaster risk management, developing and promoting policy frameworks at State and National level.

Development of Disaster Management Plans

One of the activities under the DRM programme includes, preparing disaster management plans at all levels with capacity building of stakeholders for better preparedness and response before and at the time of disaster. The Disaster Management plans being developed at each subsequent administrative level starting from the village community through to state government are as follows :

1. Village Disaster Management Plan (VDMP)
2. Taluka Disaster Management Plan (TDMP)
3. District Disaster Management Plan (DDMP)

Village Disaster Management Plan (VDMP)

A Village Disaster Management Committee (VDMC) is being formed in each programme village. This will include members ranging from PRI representatives to talathis to school teachers and community leaders. The village DM plan will be developed by the community through these nominated representatives. These plans at different levels will have information on resources, vulnerable elements and standard operating procedures (SOPs) for response groups. Each of these plans will be ratified by the concerned gram sabha.

Response Groups

Apart from these there will be different kinds of officially mandated response groups that will play an active role in the pre, during and post disaster scenario. At the community level various Village Task Forces (VTFs) will be formed, aimed at preventing a high degree of loss of life, livelihoods and property in case of a cyclone, flood or earthquake. The members will be selected from among motivated and responsible volunteers with the relevant skills sets who can implement and supervise the activities of the DM plan for example members of village youth clubs, female members of the village self help groups, literate youth of the village, school teachers, health workers, and so on who are nominated by the local community. Each group will have between three and seven members, depending on the requirement of the specific village community. The following are some of the subjects areas for task force formation :

- Early warning and communication
- Evacuation and Temporary Shelter Management
- Search and Rescue
- Health and First Aid
- Relief Co-ordination
- Water and Sanitation
- Permanent Shelter

Methodology for Village Disaster Management Plan Development

- Firstly, a cluster meeting of 10-12 villages is called. The respective village is represented by the Sarpanch/ward panch, talathi, ANM, Gram sevak, School teacher, 2-3 natural leaders and any CBO representative if available. These meetings are being organized to sensitize these village representatives to the DRM program.
- Secondly, a gram sabha is held in each of the selected villages to explain the programme elements to the community and to create a receptive environment for acceptance of the

program. The village leaders/functionaries who had been oriented in the cluster meetings will take a lead in organizing and executing the same, thus ensuring greater and faster acceptance.

- Thirdly, a Participatory Rural Apprarisal (PRA) exercise focusing on DM aspects is held in the village. The outputs generated from the PRA (which take the form of mapping exercises and interactive group discussions) provide the numbers, attributes, and lists for the VDMP.
- Fourthly, the VDMC and VTF members are nominated through a general consensus from the community and their roles and responsibilities discussed. Their standard operating procedures are then formally incorporated into the plan for that particular village.
- Fifthly, the VDMP is drafted by the community with the help of the government's extension officer/UN volunteers. It contains the physical, socio-economic, demographic and infrastructural profile of the village, a resource inventory, hazard, risk, vulnerability and capacity analysis and a list of standard operating procedures.
- Every six months a mock drill will be organized to ascertain the level of preparedness of the VAC / VTF members/ community and plans will be subsequently updated.

Taluka Disaster Management Plan (TDMP)

A Taluka Disaster Management Committee will be formed in each of the programme talukas that will have the responsibility of preparing the TDMP for their respective taluka.

This committee will comprise the respective nodal officers in the line departments and administrative offices who will prepare the TDMP with the help of their subsequent team members. All the village level data that form a part of the VDMPs in the villages of a particular taluka, are aggregated to generate the baseline for the Taluka DM Plan. In addition to this, taluka level secondary information is incorporated. A risk assessment that includes vulnerability/hazard mapping and capacity assessment for the taluka and a set of government departmental SOPs are incorporated to form the final Taluka Disaster Management Plan.

Response groups at the taluka level will take the form of line departments with their respective SOPs for emergency preparedness and response. The following are some of the nodal functionaries at this level

- Mamlatdar
- TDO
- Police
- Dy Ex Engineer from R&B, Irrigation, GEB and GWSSB
- Local municipality
- Home Guards and Civil Defense Units
- NGOs and Community Based Organizations

The detailed roles and responsibilities of these representative departments will be outlined in the plan that will explain what each of the nodal officers in these departments is expected to do in the different phases of the DM cycle.

District Disaster Management Plan (DDMP) will follow a similar methodology as the TDMP at a progressively higher level using aggregated taluka level data. The District plan will include all departmental plans made by the concerned department entrusted with response and planning at district level; It will also outline the area of inter-agency coordination. It will also outline the area of inter-agency coordination. The response groups at this level will comprise the officers of the nodal departments in the district administration.

Plan	Development
VDMP	VDMP Village Disaster Management Committee (VDMC) include PRI representatives, talathis, school teachers, other village level functionaries and community leaders
TDMP	Taluka Disaster Management Committee (TDMC) include the respective nodal officers in the line departments, administrative offices in the taluka and PRI members
DDMP	District Disaster Management Committee (DDMC) include respective nodal officers in the line departments, administrative offices in the district and PRI members

Capacity Building

At the village level, the task force groups and the Village Disaster Management Committee which co-ordinates, monitors and guides the activities of the above mentioned groups are trained in the basics of disaster management, the importance of community based disaster preparedness for risk reduction and in the above mentioned task force subject areas. At the taluka and district levels, a core team of trainers will be formed from among the government functionaries, representatives of training organizations at these levels and voluntary groups. For instance, a Health and First Aid training team would constitute some of the PHC doctors, Red Cross members and representatives from the district level training institutions of the Heath and Family Welfare. Dept. At the state level ATIs,. representatives of the nodal 'departments, technical trainers such as the professors of engineering and medical colleges will be consulted in the specifics of the DRM programme and opportunities for its integration into their help will be identified. The SPIPA and later the up-coming Gujarat State Disaster Management Institute in Gandhinagar will play a lead role in this area.

A whole range Manuals are being developed for use in capacity building including:

(i) Subject specific manuals

(ii) Hazard specific manuals

(iii) Training manuals

The manuals will essentially cater for two broad levels, one, for the village level DM functionaries and the other for the stakeholders at the taluka and higher tiers. They aim to provide a self help guide for individuals playing a role in the emergency preparedness and response scenarios. A DRM implementation handbook will also be developed that will serve as an umbrella to-do document. As mentioned earlier the SOPs for all the implementing stake-I holders will be developed and distributed.

Awareness Generation

In order to ensure that information related to all aspects of disaster management is communicated to the stakeholders at all levels, a multi-pronged communication strategy has been developed reaching out to the individual, local communities, and the general public at large. At individual level, materials such as hazard and subject specific Do's and Don'ts checklists and information booklets are being developed. At the community level posters, season specific disaster calendars, leaflets, wall paintings, rallies, mass meetings and competitions such as essay-writing, debate and drawing among school students, are envisaged. The four main mass media vehicles of television, radio, print press and the internet are going to be employed in order to ensure that DM becomes a core concern from the individual and household to the national level. The corporate sector will also be made more aware by focusing on man-made hazards like industrial and chemical disasters.

Role of Information Technology

The DRM program requires an ongoing flow of substantial information. Data generated for the plans have been thought to be captured on a uniform format for easy compilation, collation, analysis and accessibility with the assistance of the state-wide IT network of the government entitled Gujarat State Wide Area Network (GSWAN). (GSWAN) is a state-of-the art wide area network catering for government officials at State, District and Taluka levels. The State Disaster Resource Network (SDRN), a system that is being enabled by GSWAN includes some of the following information: hazards faced, data on vulnerable elements, information on past disasters, inventory of resources available, task forces and disaster management committees at various administrative levels with their emergency contact numbers. The system will allow for the creation of and updating of the Village, Taluka, District and State Disaster Management Plans once the data has been collected at the village level, entered into the SDRN system at the taluka/district level and aggregated with secondary data pertaining to the specific taluka or district. This decentralized system presents many advantages such as the easy availability of plans at levels, minimum duplication and time saving and finally the visual data reports generation that assists in gap analysis and resource mobilization.

Sustainability of the Programme as a Community led Initiative

The mainstreaming of DM into the development process is envisaged by the institutional arrangements put into place. The mock drills will become a regular feature of the village and departmental calendars. In order to ensure that the task force groups remain active, interested and motivated and are able to use the skills they have acquired from the DRM program, it is imperative that they are involved in the ongoing development programs of government and NGOs in the villages. Finally, DM plans will be incorporated into the development plan of the respective units drawn out by their concerned planning boards.

It is envisaged that the above activities combined with specialized support to the Ministry of Home Affairs, Government of India and the State Government for ensuring administrative, institutional, financial and legal mechanism for disaster risk management will result: in the reduction of expenditure on disaster relief, the mitigation of the impact of future disasters and protection of development gains and finally disaster risk considerations mainstreamed into development.. In the long term, one of the key indicators of success could be the extent to which the communities inculcate disaster preparedness as a societal trait.